PENGUINS!

Based on the series by John Tartaglia
Written by Margie Markarian

Houghton Mifflin Harcourt
Boston New York

Photo credits:
Paul Souders: 5 (top), 6, 7 (bottom), 11–17, 19, 20
Joseph C. Davola: 5 (bottom)
Nick Dale: 7 (top)
Ronnie Chua: 8
Richard Robinson: 9, 10
Dean Bertoncelj: 18

ISBN: 978-0-358-05611-9 paper over board
ISBN: 978-0-358-05612-6 paperback

hmhbooks.com

Printed in China
SCP 10 9 8 7 6 5 4 3 2 1
4500770652

Welcome to Reeftown!

I'm Splash. I love to explore!

I'm Bubbles. I love adventure!

This is Dunk and Ripple. We're the Reeftown Rangers. Today, we're going to get to know some penguins.

Did you know that there are 17 different kinds of penguins? Penguins are birds, but they can't fly. They are super swimmers!

Gentoo penguins are the fastest! Ready! Set! Go!

Emperor penguins are the biggest. Their chests are puffier than mine!

emperor penguin

Fairy penguins are the smallest. I love their blue feathers.

fairy penguin

1 Habitat: Where Penguins Plunge and Play

Penguins live in the southern half of the Earth, below the **equator**. →

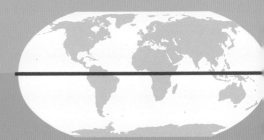

I swam with chinstrap penguins in the chilly waters of Antarctica.

chinstrap penguin

Adélie penguins

I saw a family of Adélie penguins playing on an **iceberg**!

king penguins

I hope I meet a king penguin before my fins freeze!

Some penguins live in places that are warm.
They live on beaches and in forests.

Humboldt penguins swim in the waters off the coast of South America.

Humboldt penguins

Fiordland penguin

Fiordland penguins hide in the **rainforests** of New Zealand!

Physical Features:
Dressed to Impress!

Like other birds, penguins have feathers. Their feathers protect them from cold air and water.

That penguin has a style that really rocks!

erect-crested penguins

Most adult penguins **molt** and get new feathers each year. Baby penguins have soft, fluffy feathers. When they molt, they get smooth grown-up feathers!

emperor penguin chicks

I wonder if penguins feel *chilly* and *silly* without their feathers!

molting penguin

Penguins flap their wings to swim in the water.

king penguins

Penguins spread their wings
for balance when they walk.
They tuck them in to stay
warm when they stand.

Adélie and gentoo penguins

Penguins look
like people in
fancy tuxedos!

Penguins have webbed feet with sharp claws.
Their feet help them paddle in the water.
They grip ice, snow, and rocks on land.

emperor penguin

It's funny how penguins
waddle as they walk!

Behavior: Chilling Out Together!

Penguins live in a **colony**. Many stay with the same mate for life.

emperor penguins

A penguin returns to the place it was born to lay eggs. Some penguins march a long way to get to their **rookery**.

Wow! It's crowded, noisy, and smelly when penguins get together!

Male gentoos give their mates a pretty pebble when they want to have a family. Sweet!

Some penguins build nests using rocks and pebbles. Others use grass and twigs.
Some dig holes or live in holes between rocks.

Most penguin parents take turns keeping the eggs warm.

gentoo penguins

emperor penguins

Emperor penguins don't use nests. While emperor moms look for food, emperor dads keep their egg warm in a pouch on their belly!

He looks cold standing out there. I'm glad I don't have feet!

Penguins "talk" to each other.
Fairy penguins sing.
Gentoo penguins honk.
African penguins bray like donkeys!

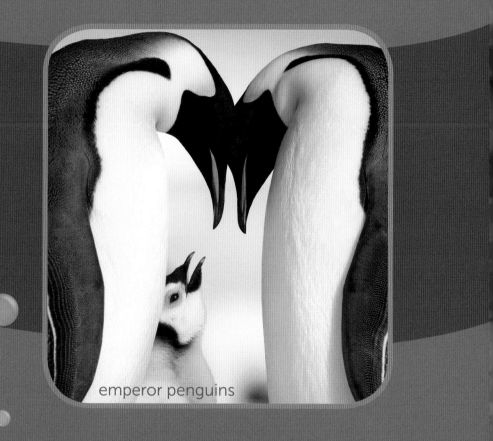

emperor penguins

All penguins bow and flap to say hello.
Chicks tap on their parents' beaks to
say "Feed me!"

Penguins eat all types of fish.

Their hooked beaks catch
krill, squid, crabs, and other seafood.

Humboldt penguin

I think I'll keep
my distance
from penguins.

Penguins do not have teeth.
They have sharp barbs on their tongue.
The barbs help them catch slippery fish.

gentoo penguin

king penguins

Penguin parents help their newborns eat.

They spit up bits of their own food
and put it into their chick's mouth!

I love my papa,
but I'm glad I
can feed myself!

Penguins are loyal, proud, and playful.

But penguins face dangers too,
like pollution, **predators**, and changing weather.
Some types of penguins could become **extinct**.

Learning and sharing what you know about
penguins are good ways to help penguins!

We love
making friends
in the ocean!

True or False?
Test Your Penguin Smarts!!

1. Penguins only live in cold places.   True   False

2. Penguins use their wings to fly in the sky.   True   False

3. Fairy penguins have blue feathers.   True   False

4. Penguins live in colonies with other penguins.   True   False

5. Emperor penguin moms are in charge of keeping eggs warm until they hatch.   True   False

6. Penguin babies are called chicks.   True   False

Glossary

colony – A group of animals that live together.

equator – The invisible line that separates the Northern Hemisphere (top half) of Earth from the Southern Hemisphere (bottom half).

extinct – A species of animal that has completely died out is extinct. Extinction is natural, but pollution, hunting, or other dangers can cause animals to become extinct more quickly.

iceberg – A large piece of floating ice.

molt – The process of shedding old feathers and growing new ones.

predator – An animal that eats other animals.

rainforests – Tropical forests with many different types of plants and animals.

rookery – A breeding ground where penguins live, make nests, lay eggs, and raise their young.